FOCUS ON™ Biographies

Bill Gates

Jennifer Strand

KAEDEN™
PUBLISHING

Kaeden.com
(800) 890-7323

ISBN: 978-1-63584-576-1

Cover Photo: Ammar Abd Rabbo/Sipa USA/AP Images
Interior Photos: Ammar Abd Rabbo/Sipa USA/AP Images, 1; Paul Sakuma/AP Images, 4–5, 16; Seth Poppel/Yearbook Library, 6, 7; akg-images/Paul Almasy/Newscom, 8; iStockphoto, 9; Ann E. Yow-Dyson/Getty Images, 10–11; AP Images, 13; Marty Lederhandler/AP Images, 14; David Cooper/Toronto Star/Getty Images, 15; Taina Sohlman/Shutterstock Images, 17; John Stillwell/PA Wire URN:12584895/Press Association/AP Images, 18; Dave Thompson/PA Wire URN:25329086/Press Association/AP Images, 19

Title: FOCUS ON Biographies: Bill Gates
Author: Jennifer Strand
Editor: Emily Temple
Series Designer: Madeline Berger
Art Direction: Dorothy Toth

Printed in Guangzhou, China
NOR/0518/CA21800519

First edition: 2018

Table of Contents

Introduction

Bill Gates helped found Microsoft. This company helped make some of the first personal computers. Gates is the richest person in the world.

Early Life

Bill was born on October 28, 1955.

Computers were new when Bill was in school. Bill's school rented one. He used it to write **programs**.

Not many people owned computers at this time. Most computers were very large.

A few computers could be kept in homes. But they did not have keyboards or screens.

Leader

Gates and Paul Allen made a program. It helped computers do more.

In 1975 they started a technology company. They called it Microsoft.

History Maker

A company called IBM wanted to make a personal computer. It hired Microsoft to help. Gates helped design the computer's **software**.

Other companies bought his software, too.

In 1985 Gates helped make an **operating system**. It was called Windows.

Windows was **popular**.
Gates became a billionaire.

Legacy

Gates helped make computers part of everyday life. Many people use his computers and software.

Microsoft became the largest technology company in the world.

Gates uses his success to help people. He gives away billions of dollars.

He works to fight **poverty** and diseases.

Quick Stats

Bill Gates

Born: October 28, 1955

Birthplace: Seattle, Washington

Wife: Melinda Ann French

Known For: Gates helped start Microsoft. He is the richest person in the world.

Key Dates

1955: William Henry Gates III is born on October 28.

1975: Bill Gates and Paul Allen start Microsoft.

1981: Microsoft helps IBM make its first personal computer.

1985: Microsoft introduces the Windows operating system.

1994: Gates becomes the wealthiest person in America.

2000: Gates and his wife start the Bill & Melinda Gates Foundation.

Glossary

operating system – the main program that controls a computer.

popular – liked or enjoyed by many people.

poverty – when people do not have enough money to live well.

program – a set of instructions that tells a computer what to do.

software – the programs that run on a computer and allow it to do certain things.

Index

FOCUS ON™ Biographies

BOOKS IN THIS SERIES

Muhammad Ali
Neil Armstrong
Beyoncé
Jeff Bezos
George Washington Carver
Roberto Clemente
Marie Curie
Walt Disney
Thomas Edison
Henry Ford
Bill Gates
Jane Goodall
Mae Jemison
Steve Jobs
Bruno Mars
Barack Obama
Katy Perry
Eleanor Roosevelt
J.K. Rowling
Charles Schulz
Maurice Sendak
Dr. Seuss
Taylor Swift
Justin Timberlake
Pharrell Williams
Wright Brothers
Malala Yousafzai